BABY ANIMALS

BABY GRAY WOLVES

by Julie Murray

Cody Koala

An Imprint of Pop!
popbooksonline.com

Hello! My name is
Cody Koala

This book is filled with videos, puzzles, games, and more! Scan the QR codes* while you read, or visit the website below to make this book pop.

*Scanning QR codes requires a web-enabled smart device with a QR code reader app and a camera.

abdobooks.com

Published by Pop!, a division of ABDO, PO Box 398166, Minneapolis, Minnesota 55439. Copyright ©2024 by Abdo Consulting Group, Inc. International copyrights reserved in all countries. No part of this book may be reproduced in any form without written permission from the publisher. Cody Koala™ is a trademark and logo of Pop!.

Printed in the United States of America, North Mankato, Minnesota.
102023
012024

THIS BOOK CONTAINS
RECYCLED MATERIALS

Cover Photo: Getty Images
Interior Photos: Getty Images; Shutterstock Images
Editors: Elizabeth Andrews and Grace Hansen
Series Designer: Candice Keimig

Library of Congress Control Number: 2023938791

Publisher's Cataloging-in-Publication Data
Names: Murray, Julie, author.
Title: Baby gray wolves / by Julie Murray
Description: Minneapolis, Minnesota : Pop!, 2024 | Series: Baby animals | Includes online resources and index.
Identifiers: ISBN 9781098245221 (lib. bdg.) | ISBN 9781098245788 (ebook)
Subjects: LCSH: Animal babies--Juvenile literature. | Animals--Infancy--Juvenile literature. | Wolves--Juvenile literature. | Timber wolf--Juvenile literature. | Gray wolf--Behavior--Juvenile literature.
Classification: DDC 591.39--dc23

Table of Contents

Chapter 1

Born in a Den

Baby gray wolves are called pups. Female wolves can have up to ten pups at a time. The pups are born in a den. They are safe and warm.

A pup weighs about one pound (0.45kg) at birth.

Watch a video here!

Pups are born deaf and
blind. They often have dark
fur and blue eyes. Their eyes

change to a golden color as they grow. Their fur color can change too.

Chapter 2

From Milk to Meat

Pups stay in the den for their first few months of life. They drink their mother's milk and grow. They eat four to five times a day.

Learn more here!

Soon, they are strong enough to move around in the den. Adult wolves begin to feed the pups **regurgitated** meat. This helps them gain strength quickly.

Chapter 3

Growing Up

When the pups are ready, they leave the den to explore. The other pack members watch over the pups. The pups are playful. They like to chase and wrestle one another.

Explore links here!

When the pups are strong
enough, they move to an
open area where the pack

sleeps, plays, and eats
together. The pups learn how
to live in a pack and hunt.

Adult gray wolves have
fur that can be gray, brown,
black, or tan. Some can even
be white! They have a **bushy**
tail and sharp teeth.

An adult male gray
wolf can weigh
145 pounds (65.8kg)!

Chapter 4

Living in a Pack

Gray wolves live in the Northern **Hemisphere** in groups called packs. A pack can have up to 15 members. An **alpha** male and female are the leaders of a pack.

Where Gray Wolves Live

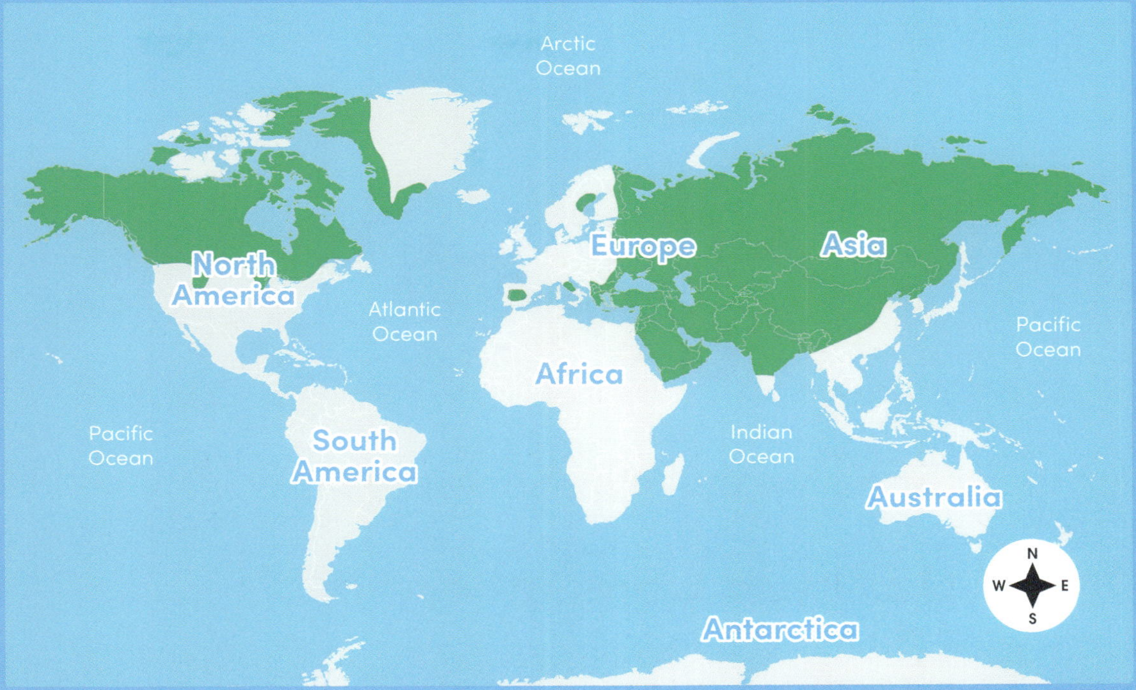

All the adults in the pack help take care of the pups.

Complete an activity here!

Wolves move around to find food. They can live in many different **environments**. Wolves hunt moose,

deer, and elk. They also eat smaller animals. The pack works together to hunt larger animals.

Making Connections

Text-to-Self

Imagine you are on a hike and see a wolf. Are you afraid? Why or why not?

Text-to-Text

Have you read another book or watched a movie about wolves? What did you learn about wolves from the movie or book?

Text-to-World

Gray wolves hunt in packs. Can you think of any other animals that hunt in groups?

Glossary

alpha – the dominant, or leading, animal in a group.

bushy – thick and fluffy.

environment – all the surroundings that affect the growth and well-being of a living thing.

hemisphere – either of the two halves of the earth. A hemisphere is formed by dividing the earth into the Northern and Southern Hemispheres at the equator, or into the Eastern and Western Hemispheres at the meridian.

regurgitate – to bring swallowed food up again to the mouth, often in order to feed young.

Index

Online Resources

popbooksonline.com

Thanks for reading this Cody Koala book!

This book is filled with videos, puzzles, games, and more! Scan the QR codes* while you read, or visit the website below to make this book pop.

popbooksonline.com/baby-wolves

*Scanning QR codes requires a web-enabled smart device with a QR code reader app and a camera.